Beyond The Infinite

By

Robert Charles Lewis

I0469342

This book is dedicated to:

1. Our Creator, who I call my FATHER.

2. My 3 children, and their mother.

3. My parents, Paul and Betty Lewis.

4. My personal Lord, Saviour, King and FRIEND, namely, JESUS CHRIST OF NAZARETH, who taught me and showed me LOVE thru examples, as written in His Holy WORD.

Foreword

This book is written in the hope that new ways of thinking can lead to a better understanding of nature, which will lead to benefits for all mankind.

This hope is also my prayer of blessing for all.

Sincerely, in the name of Jesus Christ,

Robert "Chuck" Lewis

Gravity

A form of energy

that flows to mass

and will move an object

as far as I see

doesn't run out of gas

and it's a bit abstract.

Incoming gravity

must collide

then perhaps its energy

that we could ride.

Gravity 2

Gravity must be visible

in some way

and also tangible

this I say.

It must be more than

a one-way flow

I'm sure that I can

somehow this show.

Gravity 3

When particles of gravity

collide head-on

heat must be

shed no con.

But gravity is not heat

nor electricity

it gains by friction

a temperature that's neat

thus it goes thru eternity

as an invisible action.

Gravity 4

From all directions

thru outer space

I'm sure gravity flows

with countless interactions

and it is like a race

this is plain as my nose.

Objects can gain speed

thru the slingshot effect

tangible gravity they need

and a redirect.

Gravity 5

By using gravity's energy

of celestial bodies

the universe expands

for galaxies I see

slingshot with ease

off the heavenly bands.

Gravity 6

Riding thru the cosmos

powered by gravity

for all modern science knows

it could be the dark energy.

Faster than light

gravity must be

an energy for sight

to a learned entity.

And my next thought here

is gravity in an atmosphere.

Gravity 7

An atom has gravity

and an atmosphere

split an atom for study

but don't stand near.

Now where in space

would the energy race

of an atom split there

without any atmosphere?

Gravity would split

and thus exit

and that is it.

Gravity 8

Gravity is so strong

in some locales

I am not wrong

it tips the scales

of a light source

to not permit

so of course

light won't emit.

As for the light

its inner glow

builds in might

then steals the show

for light breaks thru

as "Old Faithful" too.

(Gravity causes pulsars).

Gravity 9

Containing all creation

in the universe

are flows of gravitation

limiting the course

even of galaxies

as they rotate

understand with ease

at your dinner plate.

No form of energy

can pass thru it

as I write truly

outside I sit.

All of nature

is from our Creator.

Gravity 10

A finite universe

is what we live in

and gravity of course

is a surrounding skin.

As heavenly bodies

move farther away

give me a chance please

and hear what I say.

Gravity will rein them in

with many a lasso

and they will begin

moving closer I know.

Gravity 11

As our universe seems

to expand it shows

our Maker's gleams

in multiple glows.

A Doppler effect

transforms energy

now I elect

to explain simply.

When vehicles go away

from where we be

their tail lights are red

both night and day

but drawing near we see

white light instead.

Gravity 12

A summary of gravity

this shall be.

As a unit of mass

its energy can vary

for it absorbs all

energies of any class

and thus it can carry

a very large total.

As a stationary force

it is frictional

and can give off heat

and even light of course

also it's magnetical

so let gravity be a treat.

Gravity 13

I thank you FATHER above

for your constant love

and how you made creation

for every earthly nation.

Your laws of the universe

have set bounds of course

for we your children

who put them to test often.

But you are above and free

of your laws of gravity

FATHER, show me how

to rule gravity now.

In Jesus Christ's name I pray

answer my prayer today.

Harnessing Gravity

Gravity is energy-containing matter which when properly understood can be harnessed and utilized for the benefit of mankind.

Gravity exists in all the universe, however, it has varying amounts of effect and influence from place to place within outer space.

Gravity can both flow thru space and also be static in space.

Gravity is mass that exerts force on other mass. Particles of gravity can attach to other forms of mass to cause them to either accelerate or decelerate.

Gravity flows in all directions, both linear and curvilinear .

Gravity alone, in sufficient quantities, has the energy to "split an atom".

Gravity alone can "trap" gravity, as well as trap light and other forces of nature.

The speed of gravity is not a constant thru-out the universe. For example, static gravity, where currently it is thought to be an area of no gravity, causes friction to objects moving thru space and this friction limits the speed, for now, of say, a space vehicle or probe that is man-made. Even the speed of light is restricted.

The energy of light can be diminished to the point of being extinguished, just as a meteorite is burned up entering the earth's atmosphere. The remains of light extinguished would be dark matter and also dark energy.

All matter and energy have gravity.

Some "black holes" in space could actually be huge planets or groups of planets with enough gravitational influence to cause galaxies to spin around them. Also, since planets don't emit light, their gravity could be great enough to not allow reflected light to escape, thus their dark appearance.

Considering the fact that gravity exists thru-out the entire universe, it may be true that it is a needed medium for light to travel. As an example, sound travels faster near the earth than at high altitudes. So possibly light needs gravity to travel, plus also light has gravity of itself. I personally believe that light changes speed in space depending on gravitational influences.

A key to harnessing gravity is to discover what sets it in motion, and also how it attaches its influences to light and other forces of matter.

Gravity, like light, must give off heat in certain situations.

Gravity, approaching the earth for example, will arrive from all directions. Some gravity particles will hit the earth perpendicularly and some on angles, for gravity itself will curve. Some particles of gravity will circle the earth perpetually, forming a gravitational shield. Some gravity will be "lensed" away from the earth in a curved direction and thru-out space travel a zigzag course as it is lensed repeatedly by other objects in the universe.

The circling gravitational shield can set the limit for an atmosphere's boundary with outer space. Also, circling gravity forces can spin celestial bodies.

Gravity thru-out space can have nearly infinite amounts of various strengths of effect.

Gravity colliding with itself can form pockets in space where gravity's strength is almost completely neutralized.

In other areas of space, gravitational flows can join together and form whirlwinds of gravity. These could be easily strong enough to act as an engine and to rotate galaxies around each other. These whirlwinds of gravity would be as invisible gears that, as an example, make a clock keep time. Also, a whirlwind of gravity around binary stars would build up pressure until, and causing, intermittent emissions of light.

Gravity somehow penetrates the earth and exits again, perhaps as an altered form of matter. The same amount that enters also exits. It may be subatomic particles that have a reversed attraction to mass and thus they are repelled. Incoming gravity particles might simply (by collisions with other particles of gravity already present) bounce out the particles ahead of them.

Another thing to think about is that spikes of gravitational forces may hit the earth, causing mayhem such as earthquakes, tornadoes, and such.

Gravity can probably be detected exiting an object before entering it.

Gravity orbits mass in a flow based on the center of mass, or centers, of objects of mass. For example, gravity would be similar to the skin of an apple or the peel of an orange.

Basically, gravity is a form of dark matter. However, with proper technology, it can be detected and viewed with equipment yet to be made.

Summing up, to harness gravity will take major changes in contemplations of its nature, and dedicated people to realize that it can be utilized in beneficial ways. Researchers can certainly fit new thoughts into their equations and I hope that it will be soon.

Summary Of Gravity

1. Gravity is the outer limit of the universe.

2. The universe is not expanding due to gravity's influences.

3. Gravity will not allow light to exit the universe.

4. The universe is finite in size due to gravity, but outside the universe exists infinite space.

5. Just as gravity flows around each individual object in the universe, whether an iota, atom, or galaxy, it also flows around the universe as a whole, namely, "perimeter" gravity.

6. "Stationary" gravity exists that acts as a solid and thereby is a conduit, a medium, for energy and mass flows.

7."Frictional" gravity can be either moving or non-moving gravity, which causes heat emissions, light emissions, plus much more.

8. The universe expands to a point, but not past a fixed point, then contracts to a fixed central point (each point is to be considered a set of points). These limits are controlled by gravitational influences.

9. Gravity can cause the Doppler effect on all forms of mass and energy.

10. Gravity can be: A. Whirlwind.

                    B. Evaporative.

                    C. Whiplash.

                    D. Spectral.

                    E. Condensive.

11. Gravity can pull matter, energy (and light), thru a point in space, creating what we call a "Black Hole", which actually is only a viewpoint from humanity's position in space, however, at an opposite spatial viewpoint the "Black Hole" would be seen as a dazzling array of lights and energies and masses, and even be a galaxy of any sort from, say, "ancient" to "newly born", or a heavenly body of any form.

12. Because all the universe is filled with gravity, then any movement is instantly expressed in some way, shape, or form, at all the points in the universe, including the farthest, and because gravity is basically a "moveable solid", therefore any movement within gravity's realm has an instant offset to the limits of the universe, meaning that instantaneous space travel is humanly possible.

13. As gravity encloses the entire universe (in many ways) and light cannot escape the universe, then, to me, this is proof that our universe is 100% enclosed in light, meaning that the outer perimeter of the universe is light.

<u>Time</u>

1. Time, at the center of the earth, rotates around a point of nothingness, therefore time, at that point, does not exist.

2. Time is, on earth, simply the "physical, positional relationship between the sun and the earth."

3. Time is only a measurement of movement.

4. Where there is no movement there is no time.

5. Since all the universe contains stationary (static) gravity, gravity itself is timeless.

6. The "outer darkness" surrounding our entire universe is also timeless.

7. The past and the future are ever-present.

8. Work is also a measurement of movement, requiring time.

9. Speed is a measurement of movement requiring time. To move in true zero, instant time, means that speed is only a length.

10. With time being based on motion, there can exist, relative to different objects, etc, positive time, zero time, and

negative time. This means that travel to the past and future are possible, even while remaining in zero time relative to them, meaning staying in the present while being also in the past and the future.

11. Time is relative, basically only a length from a center point.

12. Time is expressed as increments of rotation.

13. Time is a positional length, and in relationships between moving objects, time can be expressed as a physical volume.

14. To express mathematically time as negative and positive, simply pick a point to be stationary, then express oppositely moving objects as lengths, volumes, etc, in comparison to each other and the stationary fixed point. This can equate time to movement and/or to spatial relationships.

15. Gravity can warp perceptions of time thru light and its observation.

16. Time as a length, and the universe being 1 (point), time can correctly be expressed as NON-EXISTENT.

17. Time can be expressed as:

A. Volume.

B. Length.

C. Area.

D. Dimension.

E. Mindset.

18. It is always the SAME TIME :

A. On earth.

B. In the universe.

C. In outer darkness.

19. Time is a ratio between matter, etc, therefore a rotating galaxy is timed to another unit of matter simply as a ratio. For example, dog years, cat years, elephant years (lifespans) are different than human years.

The Simplicity Of Complexity

Shrouded in complexity

is always simplicity

as far as I ever see.

When two become one

new life is begun

which also is fun.

Now when there are three

it's a family

and also a tree.

The universe is one

created it begun

by Holy Trinity.

Unified Field Theory

Simply put, there does exist a stationary "field of energy" that is in all space and is timeless, supporting limitless forms of mass, energy, and dimensions. This "field of energy" that we live in currently is only the beginning for we mortals of things to come, and no person can escape eternally dwelling within its domains for ever. This is true unconditionally.

Math

1. 3/3=1 and 2/2=1 and 1/1=1 and 0/0=1 and Infinity/Infinity=1.

2. Therefore, both 0 and infinity can be exchanged for the value of 1 mathematically.

3. The infinite can be used to define the finite.

4. The finite can define the infinite.

5. Both the finite and the infinite can express zero (nothingness).

6. 0 can express the finite and the infinite.

7. Negative numbers can define both finite and infinite mathematical terms.

8. "Black Holes" can be proven thru math calculations.

9. Both the finite and the infinite can be used to redefine eternally the finite and the infinite and even negative existence.

10. Our universe can be forever expanded with new light mathematically.

Byword

As all is possible, overriding exterior influences can occur in the natural by the supernatural.

Glossary

1. Condensive Gravity: Gravity that is pulling matter, energy, etc, to a central point or set of points.

2. Evaporative Gravity: This type of gravity pulls matter, etc, away from a central point or set of points.

3. Frictional Gravity: This type causes matter to emit heat, light, other energy forms, and also restricts the speed of space objects.

4. Lensed Gravity: This refers to gravity that has its direction in space altered, such as being deflected by another flow of gravity. It, for example, is like light being focused thru either a concave or convex lens, and thusly gravity will change direction.

5. Perimeter Gravity: This term refers to the surrounding flows of gravity that exist around all mass.

6. Spectral Gravity: This type means that gravity will show different influences, similar to light passing thru a prism.

7. Stationary Gravity: This refers to gravity that is motionless.

8. Whiplash Gravity: This term describes how gravity can travel in units of length, and that the gravity at the farthest points from a stationary or slower moving point of flow, can super-

accelerate. This can be alternating flows of speed and also steady speed.

9. Whirlwind Gravity: This term refers to gravity that flows at tremendous speeds, and also that it is the type that sets the outer limit of our universe.